YOUR KNOWLEDGE HAS VALUE

Maryna Psol

Genetic knockout of Cathepsin D using zinc-finger nucleases delivered by AAV vectors

GRIN Verlag

Bibliografische Information der Deutschen Nationalbibliothek:

Die Deutsche Bibliothek verzeichnet diese Publikation in der Deutschen National-
bibliografie; detaillierte bibliografische Daten sind im Internet über http://dnb.d-
nb.de/ abrufbar.

Imprint:

Copyright © 2013 GRIN Verlag GmbH
Druck und Bindung: Books on Demand GmbH, Norderstedt Germany
ISBN: 978-3-656-52503-5

This book at GRIN:

http://www.grin.com/en/e-book/263385/genetic-knockout-of-cathepsin-d-using-
zinc-finger-nucleases-delivered-by

Genetic knockout of Cathepsin D using zinc-finger nucleases delivered by AAV vectors

Protocol of laboratory rotation

Maryna Psol

2013

ABSTRACT

Genetic engineering is known as a powerful technique for basic research and clinical applications. Recent progress in development of zinc-finger nucleases (ZFNs), which combine the DNA cleavage ability of Fok1 restriction enzyme with highly specific recognition properties of zinc-finger motifs, allows to improve efficiency and to broaden the field of use of genome editing. Here, we demonstrate our initial results in generating novel tools for Cathepsin D gene knockout in neurons based on ZFNs technology and mediated by adeno-associated virus (AAV) vectors. Pairs of AAV-ZFNs were produced and demonstrated the robust expression of nucleases in neuronal cell culture. Observed toxicity most likely was associated with heterodimerization but not homodimerization of ZFNs; cytotoxicity was greatly reduced when ZFN were provided at lower concentrations. Future studies evaluating efficiency of *Ctsd* knockout, off-target effects on molecular level and long-term outcomes *in vivo* can be performed.

INTRODUCTION

Zinc finger nucleases (ZFNs) are engineered restriction enzymes which are employed as effective and versatile tools for targeted genome editing. Each ZFN consists of two functional domains: *Fok*1 restriction enzyme for DNA cleavage and zinc fingers (ZFs) for specific DNA binding (Figure 1). After heterodimerization of a ZFNs pair in an inverted orientation, *Fok*1 can induce double-strand breaks (DSBs) between the DNA sequences, specified by ZFs motifs. Assemblies of ZFs recognize DNA in a modular fashion, where a single ZF protein interacts with a single triplet of nucleotides, thereby allowing highly specific targeting of any DNA sequence in a complex genome (Palpant NJ, 2013). DSBs dramatically increase efficiency of gene targeting by stimulating two evolutionary conserved repair mechanisms. First is homologous recombination which underlies targeted gene replacement between existing sequence and designed donor DNA. Second, non-homologous end-joining, is a rapid but error-prone mode of DNA repair, which provides targeted mutagenesis by small insertions, deletions, substitutions etc (Caroll D, 2011).

ZFNs gene targeting was successfully applied in numerous model organisms including *Drosophila melanogaster, Danio rerio, Caenorhabditis elegans, Xenopus, Arabidopsis*, rodents etc (Palpant NJ, 2013) and various human cell lines and primary cells, such as T lymphocytes, mesenchymal stromal cells, embryonic stem cells, induced pluripotent stem cells etc (Händel EM, 2012). Gene editing frequencies up to 50% were reported (Lombardo A, 2007). But frequencies of desired gene modifications vary between different cell types and developmental stages, and depend on the method of ZFNs delivery (Caroll D, 2011), which stipulates a necessity to understand the biology and to assess an efficiency of gene targeting in every system under study.

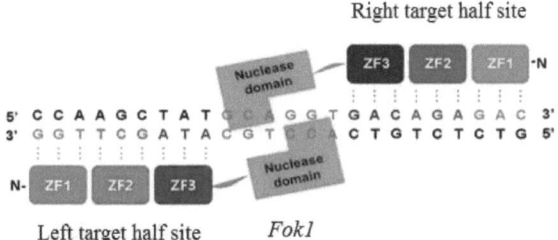

Figure 1. Schematic structure of ZFN pair bound to the target DNA (modified from http://pnabio.com/pna/ZFN.htm#)

3

Genetic engineering mediated by ZFNs is a promising approach to alleviate genetic diseases. Thus, ZFN knockout of CCR5 is applied in current HIV/AIDS clinical trial. Moreover, ZFNs technology allows developing procedures for gene manipulation in non-traditional model organisms (Caroll D, 2011). Injections of ZFNs mRNA into mouse embryo enable fast generation of genetically-engineered mouse models, which creates an alternative to more time-consuming and expensive conventional gene targeting in embryonic stem cells (Sung YH). However, genetic engineering with ZFNs raises certain ethical and safety issues. ZFNs experiments require careful assessment of the off-target effects and finding a balance between ZFN-associated toxicity and essential level of nuclease activity (Händel EM, 2012).

In current project we are aiming to generate on the basis of ZFNs technology a novel tool for Cathepsin D gene (*CTSD*) knockout. Cathepsin D is the major lysosomal aspartic protease involved in regulation of numerous proteolytic pathways (Qiao L, 2008). Deficiency of this enzyme leads to the congenital form of human neuronal ceroid lipofuscinosis (NCL) – a devastating neurodegenerative disorder, which manifests clinically with extreme neuronal loss leading to microcephaly, status epilepticus, respiratory insufficiency, astrogliosis and microglia activation, and death within hours to weeks after birth (Siintola E, 2006). *CTSD*^{null} mice develop normally until P14, but afterwards acquire CNS and intestinal pathologies followed by premature death at postnatal day 26 ± 1 (Shevtsova Z, 2010). Cathepsin D is likely to be involved in degradation of α-synuclein (Qiao L, 2008; Sevlever D, 2008) and Tau (Khurana V, 2010). *CTSD* knockout models demonstrate extensive aggregation of mentioned above proteins and prominent neurotoxicity, while overexpression of the protease has a protective effect against α–synuclein–induced neuronal loss (Qiao L, 2008). Moreover, the functions of cathepsin D are not restricted only to its enzymatic activity in lysosomes. Thus, it is involved in apoptotic pathways and acts as a mitogen, inducing progression and metastasis of certain tumors, e.g. breast cancer, and is considered to be a marker for poor prognosis. Procathepsin D is heavily secreted from various cancer cells; it supports tumor microenvironment and induces angiogenesis (Benes P, 2008; Masson O, 2010). Despite high interest of research community to Cathepsin D, many of its' specific functions and substrates are yet to be discovered.

The goal of our study was to bring together ZFNs technology with adeno-associated virus (AAV) delivery system for *CTSD* knockout in neurons. We generated the vectors with different levels of ZFNs expression and evaluated their toxicity due to homo- and heterodimerization in neuronal cell culture. This work underlies further *ex vivo* and *in vivo*

4

assessments of AAV-mediated ZFNs tools for *CTSD* ablation and their implementation in Cathepsin D research.

MATERIALS AND METHODS

Molecular cloning

Cloning procedure was simulated with the SECentral software. The constructs encoding ZFN1 and ZFN2, mur*CTSD*_pZFN1 and mur*CTSD*_pZFN2 (Figure 2), were produced by Sigma (Sigma-Aldrich, USA). As a backbone for subcloning we used the plasmid AAV-6P-NoTB-SEWB (Figure 3) (kindly provided by Dr. S.Kügler).

5-10 µg of plasmid DNA were used in each restriction digest reaction. Appropriate restriction enzymes and buffers (New England BioLabs, USA) were added to DNA and incubated for 1-2 h at the specified by manufacturer temperature. DNA fragments were separated on preparative 1% agarose gel. Desired bands were cut out and subjected to the gel extraction according to the QIAGEN kit protocol. Eluted DNA was precipitated with 3M sodium acetate and 100% ethanol in order to increase concentration and purity of the samples. Concentration of DNA was determined with analytical gel against λ-HindIII-standard. For ligation a vector and an insert DNA were added in 1:3 molar ratio and mixed with the T4 DNA ligase and T4 ligation buffer. Ligation reaction was performed for 20 min at room temperature. The product of ligation was electroporated into electrocompetent *E.coli* cells. Transformed cells were incubated in SOC++ medium (2% bactotryptone, 0.5% yeast extract, 10 mM NaCl, 10 mM KCl, 20 mM MgCl$_2$ and 2 mM glucose) for 45 min at 37°C and then plated on LB agar plates containing ampicillin (100µg/ml) for the selection of the clones. DNA plasmid extractions were performed using the QIAGEN Plasmid Mini- and Megaprep kits according to the protocol of the manufacturer. Subsequent control digestion with *SmaI*-enzyme was used to confirm a ligation of proper DNA fragments (Figure 4B).

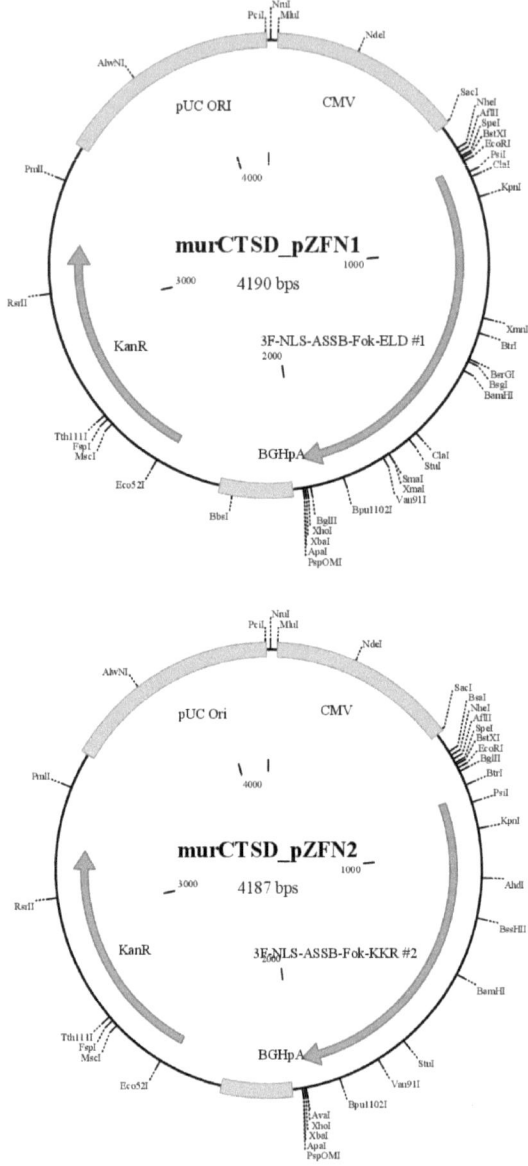

Figure 2. Schematic depiction of the original plasmids encoding ZFN1 and ZFN2. Sequences encoding zinc finger nucleases are marked as 3F-NLS-ASSB-Fok-ELD #1 and #2

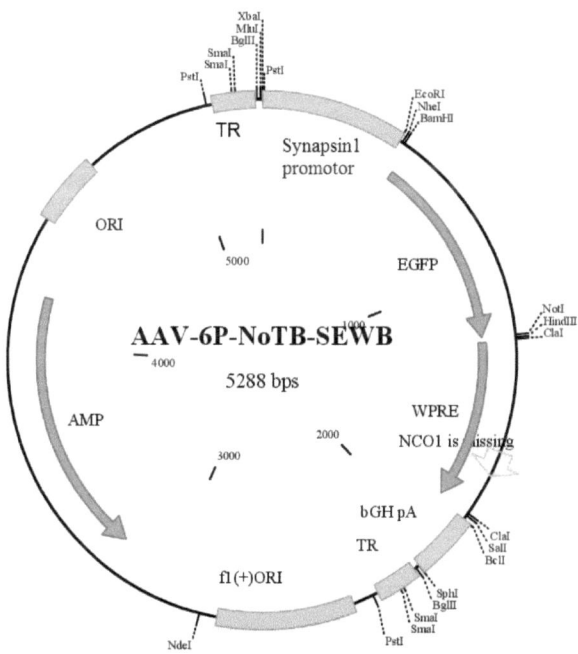

Figure 3. Schematic depiction of the plasmid AAV-6P-NoTB-SEWB: AAV – adeno-associated virus, NoTB – no transcription blockers, S – human Synapsin1 promotor, E – EGFP, W – WPRE (Woodchuk hepatitis virus Posttranscriptional Regulatory Element), B – bovine growth hormone gene polyadenilation site.

In order to construct AAV vector expressing zinc-finger nucleases we substituted EGFP in the plasmid AAV-6P-NoTB-SEWB for either ZFN1 or ZFN2. Backbone vector was cut with *EcoRI* and *NotI*, ZFN-constructs – with *EcoRI* and *PspOMI* (compatible with *NotI*). Fragments of interest were ligated as described above. The obtained constructs has the next structures: AAV-6P-NoTB-S-ZFN1-WB and AAV-6P-NoTB-S-ZFN2-WB. These both vectors contained WPRE (Woodchuk hepatitis virus Posttranscriptional Regulatory Element) which is a DNA sequence that enhances the expression of the AAV-delivered genes.

To decrease a level of expression of zinc-finger nucleases we produced the constructs without WPRE. For this the AAV-6P-noTB-SEWB was cut with *ClaI* and religated on itself. Obtained AAV-6P-noTB-SEB construct was digested with *EcoRI* and *NotI*, and murCTSD_pZFN1 and murCTSD_pZFN2 were cut with *EcoRI* and *PspOMI*. After ligation the desired fragments, the constructs AAV-6P-noTB-S-ZFN1-B and AAV-6P-noTB-S-ZFN2-

7

B were received.

Finally, we cloned the vectors without promotor which should have further reduced the expression level of ZFN. In this case expression at low level is still possible due to promotor activity of ITR regions. ZFN1 and ZFN2 were cut out with *NheI* and *PspOMI* and transferred to the AAV-6P-noTB-SEB, cut at compatible restriction sites (*XbaI* and *NotI*). As a result received vectors have such structure: AAV-6P-NoTB-ZFN1-B and AAV-6P-NoTB-ZFN2-B.

AAV production

Viral vector production and purification was done by Dr. Sebastian Kügler. All experiments were performed with AAV vectors of serotype 6. Expression cassettes included the genes for ZFN1 or ZFN2 with or without WPRE regulatory region at 3'end and under control of human Synapsin 1 promotor or without any promotor (Figure 4). As a control we used AAV6 vector encoding eGFP.

Viral vectors were propagated in HEK 293 cells. HEK cells were seeded at 1-1.2 x 10^8 cells per factory in DMEM containing 10% FCS and 1% PS. After reaching ~50% confluency cells were transfected with AAV plasmid and pDG helper plasmid with calcium-phosphate method. The cells were supplied with 2% FCS medium during the transfection and 10% FCS medium afterwards. Approximately 2 days after transfection cells were harvested using citric saline. After centrifugation cell pellets were resuspended in 20 ml Tris-buffered saline and stored at -80^0 C until further procedures. Vectors were purified by iodaxanol gradient centrifugation. AAV peak fractions were pooled with fast protein liquid chromatography (FPLC), dialyzed against PBS and stored at -80^0 C. Genome titers were determined by quantitative PCR.

Cell culture and transfection

Cell culture experiments were performed in HEK 293 cells, primary rat and mouse cortex neurons.

HEK cells were cultured at 37°C, 5% CO_2 in DMEM containing 10% FCS and 1% PS. Cells were co-transfected with plasmids containing eGFP reporter gene and plasmids with ZFN1, ZFN2 or both. Constructions with or without WPRE region were used in different series of experiments. DNA-calcium-phosphate precipitation method was used for transfection. 3μg of each plasmid were used for 1 well of a 6-well plate.

8

Rat and mouse cortex neurons were seeded at a density 250,000 cells/well in poly-L-ornithine/laminin 24-well plates. Cultures were incubated in cortex medium at 37°C, 5% CO_2 in a humidified atmosphere. Cells were transduced with 0.5-3 x 10^7 tu/well of the described above viral vectors and were harvested after 5-12 days of incubation. Arabinose was added to the medium to inhibit the growth of the glial cells. Efficiency of transduction was evaluated by the quantity of cells with and without eGFP expression and amounted to 40-70%.

Western blot

HEK 293 cells, rat and mouse neurons were lysed in the buffer containing 0.5% SDS, 1 mM DTT, 10 mM PI, 50 mM Tris. Lysates were sonicated for 10-15 sec and centrifuged with 13 000 rpm for 45 min at 4^0C, supernatant was collected, aliquoted and stored at -20^0C. The concentration of protein in the samples was determined with the bicinchoninic acid assay (BCA). Proteins were resolved with SDS-PAGE on 14% acrylamide gel and immunoblotted on nitrocellulose membrane. The following primary antibodies were used: mouse monoclonal Anti-FLAG M2 (Sigma, 1:5 000), rabbit polyclonal Anti-GFP (Clontech, 1:1 000), mouse monoclonal Anti-ß-tubulin (Sigma, 1:5 000). Secondary anti-mouse (Dianova, 1:3000) and anti-rabbit (Dianova, 1:1 000) antibodies were conjugated with HRP and signals were visualized by chemiluminescence. All Western blots were repeated 2-5 times with similar results.

RESULTS

Construction of AAV vectors

When present in cells at high concentration, ZFNs may introduce DSBs at off-target sites. However, an effective transduction of neurons requires high quantity of AAV-vector units. Due to these conditions, we produced 3 pairs of AAV-ZFN-constructs with different levels of expression, in order to search for the pair with productive nuclease activity at CTSD locus but minimal toxic effects.

In the first couple of vectors, ZFNs expression was driven under the control of human Synapsin 1 promotor, specific for neurons, and was enhanced by WPRE (Woodchuk hepatitis virus Posttranscriptional Regulatory Element). For subsequent reduction of the expression rate, WPRE sequences were deleted in the second pair of vectors. Finally, the constructs without WPRE and promotor regions were cloned, where production of ZFNs at lower level

9

should be possible due to promotor-like activity of the ITR regions. Also, the synthesis of ZFNs in the last pair is not restricted to neurons. Expression cassettes of the generated AAV vectors are depicted at Figure 4 A.

Control digestion of the AAV-ZFNs plasmids with *SmaI* restriction enzyme showed the fragments of theoretically expected size (Figure 4B). Quality of the constructs was also confirmed by sequencing.

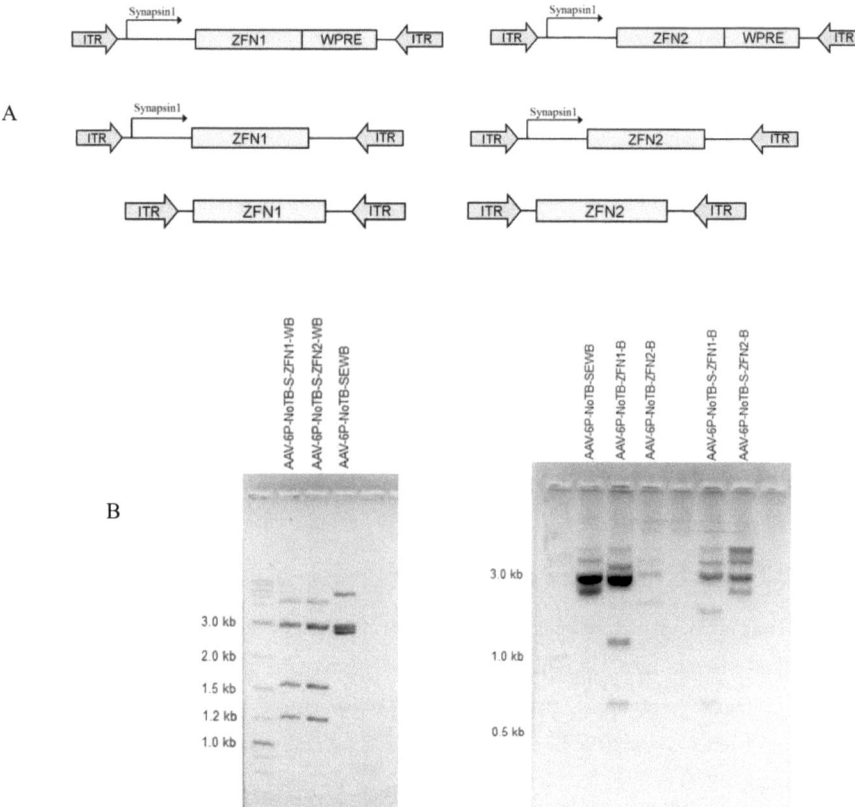

Figure 4. AAV vectors: (A) schematic structure of genetic content; (B) control digestion of the AAV-ZFNs plasmids with endonuclease *SmaI*. AAV – adeno-associated virus, NoTB – no transcription blockers, S – human Synapsin1 promotor, E – EGFP, W – WPRE, B – bovine growth hormone gene polyadenilation site.

ZFNs expression in neuronal cell culture

In order to estimate the toxicity of the AAV-Synapsin-ZFN-WPRE vectors, we transduced rat primary cortex neurons with 3 x 10^7 tu/well of either ZFN1 or ZFN2-containing viruses, or with both vectors, 1.5 x 10^7 tu/well of each. EGFP expression was used for control. Transduction efficiency, calculated as percentage of EGFP positive cells among all neurons, approximate to 50-60%. 5 days after transduction the neurons with both ZFNs revealed signs of axonal degeneration, the affected processes appeared as bean-like structures (Figure 5). Cells on other plates looked completely healthy. Neuronal cultures with one type of ZFN plus EGFP appeared less bright in comparison with cells transduced with EGFP alone. This could be explained by competitiveness of AAV vectors for transduction. The neurons with both ZFN1 and ZFN2 expressed the least amount of EGFP which could also be attributed to toxicity. Since toxic effects were observed only in the cell cultures, where both ZFNs were expressed, but not in the neurons with single ZFN, this toxicity might be the result of heterodimerization, and not homodimerization of nucleases. Although ZFNs were constructed against murine Cathepsin D gene, they were likely to produce off-target DSBs in rat genome, when supplied at high concentration.

In the rat neurons, transduced with 3 times lower vector titers (0.5 x 10^7 tu/well of each AAV-Synapsin-ZFN-WPRE) no apparent axon degeneration or decrease in EGFP expression were observed even 12 days after transduction (Figure 6).

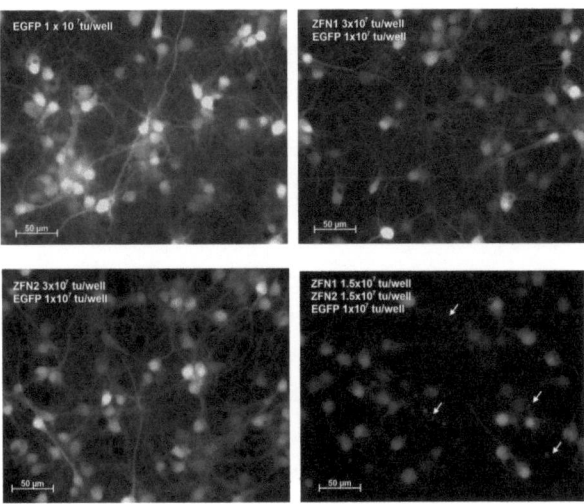

11

Figure 5. Expression of ZFNs mediated by AAV vector delivery in rat primary neuronal culture. Titers of the vectors are indicated on photos in transduction units per well (tu/well). White arrows indicate degenerated neuronal processes. Pictures were taken 5 days after transduction, with exposure time 1200 ms, x20 (Axiovert 200 M).

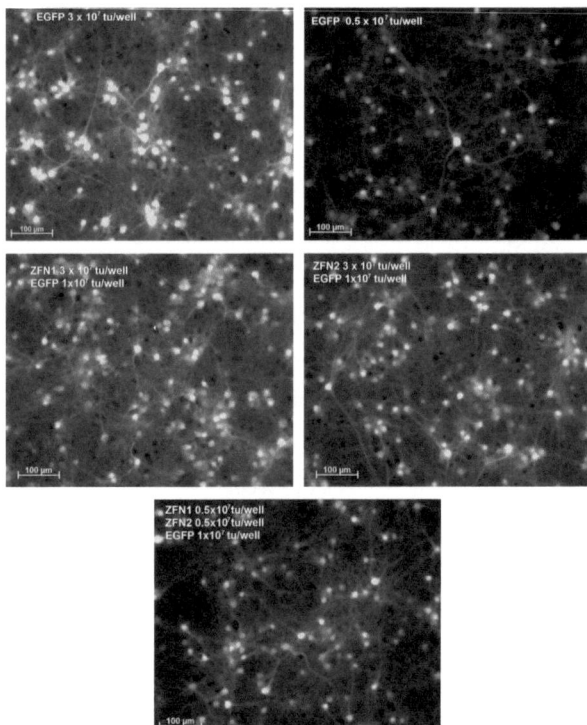

Figure 6. AAV-mediated ZFNs expression in rat primary neurons. No visible signs of axonal degeneration were detected. Titers of the vectors are indicated on photos in transduction units per well (tu/well). Pictures were taken 12 days after transduction, with exposure time 500 ms, x40 (Axiovert 200 M).

Primary cultures of mouse cortex neurons were transduced with AAV-ZFNs vectors with and without WPRE (Figure 7). Each construct was used at the titer of 1 x 10^7 tu/well. 6 days post transduction the neurons, which expressed either pair of ZFNs, manifested first signs of process degeneration, and one day later some dead cells were noticed. These toxic effects were more dramatic than in rat neurons. The cultures with WPRE-enhancement of ZFNs expression appeared less healthy than ones without WPRE. All neuronal cultures, containing

12

one type of ZFNs, demonstrated only mild signs of degeneration until the 12th day after transduction. Therefore, high concentrations of single murine *CTSD*-ZFN could be better tolerated by mouse neurons than such of ZFN pair. The titers of ZFNs pairs should be further decreased in subsequent experiments.

To demonstrate that ZFNs are expressed from the generated constructs, we performed immunoblotting analysis (Figure 9 A-C). HEK 293 cells, transfected with plasmids, as well as rat and mouse cortex neurons, transduced with AAV vectors, revealed the bands of ZFN1 at ~48 kDa and ZFN2 at ~44 kDa as expected. Also several additional bands of higher and lower molecular weight were detected. These proteins are most likely derived from the AAV-ZFN vectors and not unspecific, since they are absent in EGFP controls. ZFN2 had nearly two times higher level of expression as compared with ZFN1 in all series of experiments. The comparison of the mouse neurons, collected 7 and 12 days post transduction, reveals that ZFNs accumulates in cells with time. Removal of WPRE region allows to downregulate considerably the expression of ZFN.

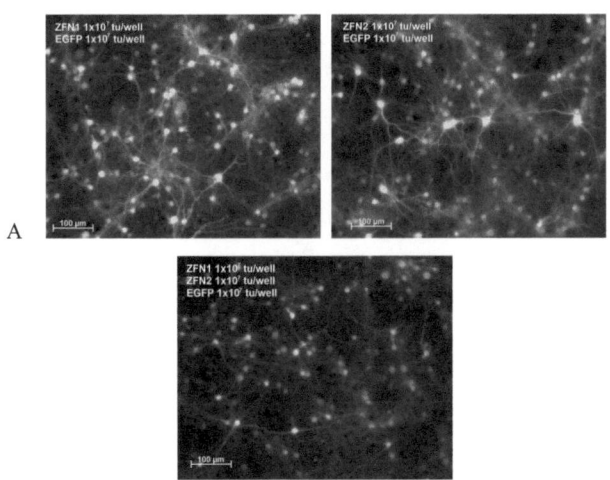

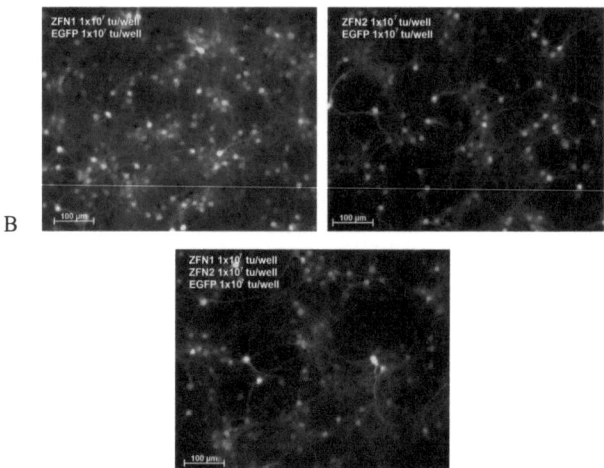

Figure 8. ZFNs expression in mouse primary cortex neurons. AAV vectors with WPRE (A) and without WPRE (B) were transduced into neurons. Vector titers are indicated on photos in transduction units per well (tu/well). Pictures were taken 7 days after transduction, with exposure time 500 ms, x40 (Axiovert 200 M).

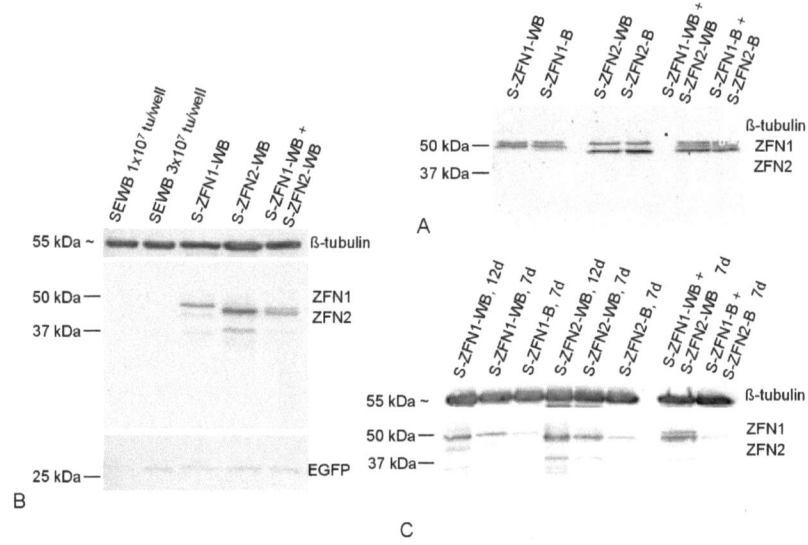

Figure 9. Representative immunoblots for detection of ZFNs in HEK 293 cells (A), primary cultures of rat cortex neuron (B), and mouse cortex neurons (C): ZFN – murine *CTSD* zinc-finger nuclease, S – human Synapsin1 promotor, E – EGFP, W – WPRE, B – bovine growth hormone gene polyadenilation site, 7d and 12d – days after transduction. The predicted molecular weights are: ZFN1 ~48 kDa, ZFN2 ~44 kDa, EGFP ~27 kDa, ß-tubulin ~55 kDa.

DISCUSSION

Our results confirm that zinc-finger nucleases, designed for deletion of murine Cathapsin D gene, can be expressed in neuronal cell culture when delivered by AAV6-vector. The level of their expression is upregulated in presence of WPRE. When *CTSD*-ZFNs were provided at high concentration, they may have provoked toxic effects leading to degeneration of neuronal processes or cell death. However, such conspicuous signs of toxicity were absent if the amount of ZFNs was low.

Efficiency and safety of ZFNs application depend on number of zinc fingers: it was shown that ZFNs with three fingers are less toxic and as active as ones with more fingers (Pruett-Miller SM, 2008), which justifies the use of 3-finger endonucleases in our project. Previous studies suggest that ZFN toxicity is induced by off-target DSBs (Pruett-Miller SM, 2008; Pruett-Miller SM, 2009). We assume that in our experiments cytotoxicity was caused by heterodimerization of ZFNs at off-target sites, because the most apparent toxic effects were observed in the cell cultures with both ZFNs. Although some degenerative features were noticed in neurons with a single type of ZFNs, they were at the same level as in EGFP control and therefore were probably caused by other factors than homodimerization of *CTSD*-ZFNs. While we evaluated toxicity of ZFNs on the cellular level, further assessment of molecular effects (e.g. visualization of DSBs) will be necessary to clear if normal physiological functioning of neurons could be maintained.

CTSD-ZFNs were accumulated in the neurons with time (Figure 9C), which probably have increased their cytotoxicity. Such toxic effects may be diminished with several strategies for regulation of the nuclease level, for instance: introduction of genetic elements that regulate transcription (e.g. TetON/TetOFF system), destabislization of ZFNs by small molecules or via fusion to ubiquitin and control of proteasome activity (Pruett-Miller SM, 2009).

Successful ablation of *CTSD* gene by ZFNs depends on AAV transfer efficiency in neurons. Although AAV2 is the most extensively studied vector, many other AAV capsid serotypes have been discovered and characterized and numerous chimeric serotypes were generated (Zincarelli C, 2008). This diversity allows choosing an AAV vector with optimal cellular tropism and transduction properties. Applied in our experiments AAV6 was previously reported to be effective in EGFP delivery to neurons, e.g. being able to transduce over 75% of dorsal root ganglion neurons (Mason MRJ, 2010). AAV6 demonstrates tropism not only to neurons, but also to glial cells, which might be useful under necessity to knockout CTSD in all CNS cells types. Alternatively, AAVs of other serotypes might be considered for ZFNs delivery in the brain. AAV9 demonstrated wide-scale neuronal transduction when administered via brain injection and, surprisingly, intravenously and intrathecally. Yet it requires relatively high vector titers for efficient gene delivery (Dayton RD, 2012; Grey SJ, 2013). Also AAV5 appeared to have high transduction rate in the neurons (Shevtsova Z, 2005; Mason MRJ, 2010).

Whether *CTSD* gene can be knocked out by generated AAV-ZFN vectors, should be still confirmed with the PCR. In case only one allele is being mutated, it will be necessary to generate a ZFNs pair against the second allele.

So far our experiments were restricted to neuronal cell culture. Since *ex vivo* findings are not always being replicated *in vivo* easily, ZFN expression and associated cytotoxicity, AAV6 transduction properties, and efficiency of Cathepsin D gene knockout should be confirmed in animal experiments.

Here, for the first time we try to knockout gene in the neurons with AAV-ZFN tools. This technique might be useful for studying Cathepsin D biology and potentially for targeting other genes in the CNS.

In conclusion, ZFNs can be robustly expressed in neuronal cell culture. Cytotoxicity, associated with heterodimerization of ZFNs pair, may be decreased by regulating ZFNs expression level. Future studies evaluating efficiency of *CTSD* knockout, off-target effects and long-term outcomes *in vivo* can be performed.

ACKNOWLEDGEMENTS

I would like to thank Dr. Sebastian Kügler for giving me the opportunity to conduct the practicum in the Viral Vector Lab in the Neurology department of the University Medical Centre Goettingen, support and valuable discussions during the project. I wish to thank techincal assistants Monika and Sonja for constant guidance in the lab and help with cell culture, virus preparation and histology, Dr. Grit Taschenberger for advices on Western blot, Dr. Julia Tereschchenko for demonstrating animal experiments, and to Dr. Muzna Zahur for working side by side with me on the project. My thanks to all people of the viral vector group for creating a friendly atmosphere.

REFERENCES

Benes P, Vetvicka V, Fusek M. Cathepsin D – many functions of one aspartic protease. *Crit Rev Oncol Hematol* 2008, 68(1): 12–28.

Carrol D. Genome engineering with zinc-finger nucleases. *Genetics* 2011, 188: 773–782.

Dayton RD, Wang DB, Klein LK. The advent of AAV9 expands applications for brain and spinal cord gene delivery. *Expert Opin Biol Ther* 2012, 12(6): 757–766.

Ellis BL, Hirsch ML, Porter SN, Samulski RJ, Porteus MH. Zinc-finger nuclease-mediated gene correction using single AAV vector transduction and enhancement by Food and Drug Administration-approved drugs. *Gene Therapy* advance online publication 2012, 1 – 8.

Gray SJ, Nagabhushan Kalburgi S, McCown TJ, Jude Samulski R. Global CNS gene delivery and evasion of anti-AAV-neutralizing antibodies by intrathecal AAV administration in non-human primates *Gene Therapy* 2013, 20: 450–459.

Händel EM, Gellhaus K, Khan K, Bednarski C, Cornu TI, Müller-Lerch F, Kotin RM, Heilbronn R, Cathomen T. Versatile and efficient genome editing in human cells by combining zink-finger nucleases with adeno-associated viral vectors. *Human Gene Therapy* 2012, 23: 321-329.

Khurana V, Elson-Schwab I, Fulga TA, Sharp KA, Loewen CA, Mulkearns E, Tyynela J, Scherzer CR, Feany MB. Lysosomal dysfunction promotes cleavage and neurotoxicity of tau in vivo. *PLoS Genet* 2010, 6 (7): e1001026.

Lombardo A, Genovese P, Beausejour CM, Colleoni S, Lee YL, Kim KA, Ando D, Urnov FD, Galli C, Gregory PD, Holmes MC, Naldini L. Gene editing in human stem cells using zinc finger nucleases and integrase-defective lentiviral vector delivery. *Nat Biotech* 2007, 25: 1298 – 1306.

Masoon O, Bach AS, Derocq D, Prébois C, Laurent-Matha V, Pattingre S, Liaudet-Coopman E. Pathophysiological functions of cathepsin D: Targeting its catalytic activity versus its protein binding activity? *Biochimie* 2010, 92(11):1635-43.

Mason MRJ, Ehlert EME, Eggers R, Pool CW, Hermening S, Huseinovic A, Timmermans E, Blits B, Verhaagen J Comparison of AAV serotypes for gene delivery to dorsal root ganglion neurons. *Molecular Therapy* 2010, 18 (4): 715-724.

Palpant NJ, Dudzinski DM Zinc-finger nucleases: looking toward translation. *Gene Therapy* 2013, 20: 121 – 127.

Pruett-Miller SM, Connelly JP, Maeder ML, Joung JK, Porteus MH. Comparison of zinc finger nucleases for use in gene targeting in mammalian cells. *Molecular Therapy* 2008, 16 (4): 707-717.

Pruett-Miller SM, Reading DW, Porter SN, Porteus MH. Attenuation of zinc finger nuclease toxicity by small-molecule regulation of protein levels. *PLoS Genet* 5 (2): e1000376.

Qiao L, Hamamichi S, Caldwell KA, Caldwell GA, Yacoubian TA, Wilson S, Xie ZL, Speake LD, Parks R, Crabtree D, Liang Q, Crimmins S, Schneider L, Uchiyama Y, Iwatsubo T, Zhou Y, Peng L, Lu YM, Standaert DG, Walls KC, Shacka JJ, Roth KA, Zhang J. Lysosomal enzyme Cathepsin D protects against alpha-synuclein aggregation and toxicity. *Molecular Brain* online publication 2008, 1: 1-17.

Sevlever D, Jiang P, Yen SHC. Cathepsin D is the main lysosomal enzyme involved in the degradation of α-synuclein and generation of its carboxyterminally truncated species. *Biochemistry* 2008, 47(36): 9678-9687.

Shevtsova Z, Garrido M, Weishaupt J, Saftig P, Bähr M, Lühder F, Kügler S. CNS-expressed cathepsin D prevents lymphopenia in a murine model of congenital neuronal ceroid lipofuscinosis. *Am J Pathol* 2010, 177 (1): 271-279.

Shevtsova Z, Malik IJM, Michel U, Bähr M, and Kügler S. Promoters and serotypes: targeting of adeno-associated virus vectors for gene transfer in the rat central nervous system in vitro and in vivo. *Exp Physiol* 2005, 90: 53-59.

Siintola E, Partanen S, Stromme P, Haapanen A, Haltia M, Maehlen J, Lehesjoki AE, Tyynela J. Cathepsin D deficiency underlies congenital human neuronal ceroid-lipofuscinosis. *Brain* 2006, 129:1438–1445.

Sung YH, Baek IJ, Seong JK, Kim JS, Lee HW. Mouse genetics: catalogue and scissors. *BMB Reports* 2012, 45(12): 686-692.

Zincarelli C, Soltys S, Rengo G, Rabinowitz GE. Analysis of AAV serotypes 1–9 mediated gene expression and tropism in mice after systemic injection *Molecular Therapy* 2008, 16 (6): 1073–1080.